Regula Pollution

Does the U.S. System Work?

by
J. Clarence Davies
and
Jan Mazurek

Resources for the Future • Washington, DC

Printed in the United States of America

Published by Resources for the Future
1616 P Street, NW, Washington, DC 20036–1400

Library of Congress Cataloging-in-Publication Data

Davies, J. Clarence.
Regulating pollution: does the U.S. system work? By J. Clarence Davies and Jan Mazurek.
     p. cm.
     ISBN 0-915707-85-3 (alk. paper)
     1.Pollution — Law and legislation—United States. 2. Environmental law—United States. I. Mazurek, Jan, 1965-   II. Title
KF3775.D34 1997
344.73'04632—dc21

97-7159
CIP

♾ The paper in this book meets the guidelines for permanence and durability of the Committee on Production Guidelines for Book Longevity of the Council on Library Resources.

This book is the product of the Center for Risk Management at Resources for the Future, J. Clarence Davies, director. The book was edited by Richard Getrich and designed by Diane Kelly, Kelly Design.

CONTENTS

PREFACE

This short report is based on a major research project undertaken by Resources for the Future's Center for Risk Management at the request of the Andrew W. Mellon Foundation. The full results of the research will be published as an RFF book in late 1997. We are issuing this report in advance of the book because we thought it important to disseminate the research results in a timely fashion and in a form that is easily accessible to policymakers.

To make this document short and readable, many of the details, implications, caveats, and other elaborations attached to our conclusions in the larger work have been omitted. Also, there are no citations or footnotes in this report. A fully annotated version of the report can be found at Resources for the Future's Web site: http://www.rff.org.

The research behind this report reflects three years of effort by a small army of diligent scholars at Resources for the Future. They include Richard Morgenstern, Mark Powell, Elizabeth Farber, Michael McGovern, Robert Hersh, Elise Annunziata, Kevin Milliman, Tom Votta, Kieran McCarthy, and Nicole Darnall. A large number of people reviewed the work and provided helpful comments. We are grateful to Paul Portney, Rich Getrich, Chris Kelaher, Allen Kneese, Mike Newman, Kate Probst, Vic Kimm, Michael McGovern, Marilyn Voigt, Pete Andrews, David Clarke, Mike Taylor, and Ruth Bell for their advice and assistance. While credit for the report belongs to a large number of people, blame should be assessed only against the two primary authors.

We are very grateful to our funding sources, the Andrew W. Mellon Foundation and the Smith Richardson Foundation. Funding also was provided by general support funds of Resources for the Future and the Center for Risk Management.

INTRODUCTION

The flagship environmental program of the Clinton administration in the 1996 election year was "Project XL"—a set of initiatives designed to circumvent the inflexibility of existing pollution control laws. If the showcase project of a pro-environmental administration is designed to help get around the environmental laws, then something is wrong with these laws and the programs that implement them. The present report is a distillation of what we know about the pollution control laws and programs in the United States and an evaluation of what is wrong and what is right about them.

This report is based on a comprehensive three-year examination of the U.S. pollution control regulatory system, the first full-scale systematic evaluation of U.S. pollution control efforts. (The full research study will be published as a book by Resources for the Future later this year.) Our goal in this study has been to evaluate the pollution control regulatory system so as to better understand the current system and to lay the groundwork for any changes that may be needed. We assume, perhaps naively, that those aspects or components of the system that are working well should not be changed, whereas those that are not working well should be changed. This report does not provide detailed recommendations for change because that is not its purpose. Its emphasis is on evaluation, not recommendation.

For all its accomplishments, we conclude that the pollution control regulatory system is deeply and fundamentally flawed. While there is no consensus about how to remedy these flaws, some agreement exists on the principles that should guide changes in pollution control and about the characteristics of a pollution control system for the next century. The United States does not need to wait for a consensus to act: to do so would be to wait forever. Failure to make the changes will be costly to the economy, to the environment, and to every citizen.

We begin the present report by describing and evaluating the legislation, the administrative decisionmaking, and the federal-state division of labor that are the main elements of the pollution control regulatory process. We then examine and evaluate the overall pollution control system using several different criteria:

- Has the system reduced pollution levels?
- Has it targeted the most important problems?
- Has it been efficient and cost-effective?
- How responsive has it been to a variety of social values (such as public involvement, nonintrusiveness, environmental justice)?
- How does it compare with the systems in other developed nations?
- How well can it deal with future problems?

Other evaluation criteria could have been used, and we settled on these only after a good deal of reflection and revision. Most previous evaluations of government programs have had a narrower focus, and therefore the criteria for evaluation have been more obvious. The goals of literacy programs or prison training programs, for example, have been relatively straightforward, so choosing criteria for evaluation has not been a major problem. Given the very broad scope of the pollution control system, the selection of the above criteria was a challenge for us; we hope that our efforts contribute to the methodology of program evaluation as well as to understanding pollution control.

The present report focuses largely, although not entirely, on federal environmental efforts. Because states and localities are responsible for much regulatory activity, our approach somewhat distorts the picture of the environmental regulatory system. We have tried to indicate those areas where nonfederal action is most important, and in the larger study we discuss state and local roles and capabilities. We have not described in detail the numerous nongovernmental groups and institutions, such as environmental advocacy organizations, corporations, and trade associations, that are an essential part of the regulatory system.

Despite these limitations, we believe that the study summarized in the present report provides a sound basis for considering the future of the pollution control regulatory system. A thorough evaluation of existing programs is the logical first step in deciding what, if anything, should be changed and what the nature of the changes should be.

LEGISLATION AND INSTITUTIONS

"Fragmented" is the adjective that comes to mind when describing the U.S. pollution control regulatory system. The system is so complex and disjointed that no one person can fully understand it. It involves hundreds of laws and agencies and thousands of interested groups. It includes some general principles (such as the establishment of uniform national standards with states being allowed to set more stringent standards), but such principles only explain some of the laws and programs. Moreover, there are numerous exceptions even to the few principles that exist.

This fragmentation is a challenge to describing and evaluating the system. We begin by describing the legal framework for pollution control. We then deal with administrative decision-making and conclude our description of the process with the federal-state division of labor.

Federal Legislation

The federal pollution control laws are the major driving force in the U.S. pollution control system. The laws dictate the system's agenda and its priorities. They specify what is to be done, when it should be done, and how it should be done. Reality often diverges from the dictates of the law, but to understand the system one must begin with the federal statutes.

There are nine major federal pollution control laws, although the number can be larger or smaller depending on one's definition of major. Three (the Clean Air Act, the Clean Water Act, and the Safe Drinking Water Act) are based on the environmental medium in which the pollution occurs; one (the Resource Conservation and Recovery Act) is focused primarily on a medium—land—but deals with other matters as well; one (the Federal Insecticide, Fungicide, and Rodenticide Act) deals with a particular set of products; one (the Toxic Substances Control Act) deals with chemicals in general; one (the Comprehensive Environmental Response, Compensation, and Liability Act) deals with accidents, spills, dump sites, and liability; and two (the National Environmental Policy Act and the Pollution Prevention Act) deal with general policy.

Added to these major acts are hundreds of minor federal laws, each with its own particular perspective on pollution control. Congress considers hundreds of environmental bills and sometimes enacts as many as twenty or thirty in each session. For example, among the bills that became law in the 103d Congress were the Federal Employees Clean Air Incentives Act, the Indian Lands Open Dump Cleanup Act, and the Ocean Pollution Reduction Act.

The major pollution control laws are quite detailed and have been getting more so. The full Clean Air Act, prior to the 1970 amendments, was only 22 pages long and contained 6 deadlines, 3 of them related to studies. The 1970 amendments are 38 pages long and contain some 12 deadlines. The 1990 amendments to the Clean Air Act are more than 300 pages long and add 162 statutory deadlines to the workload of the U.S. Environmental Protection Agency (EPA).

The statutory detail is driven primarily by a basic congressional mistrust of EPA, the agency charged with implementing the laws. Pro-environment members of Congress fear that the agency will not be ardent enough in defending the environment; members more sympathetic to business concerns fear that the agency will be too ardent. The pro-environment members (and the pro-business members) write detailed instructions into the law; the pro-business members try to ensure that the agency will not have enough resources to fully implement the laws and that the courts will have authority to second-guess any agency decisions. EPA is thus the focus of an odd and intricate system of checks-and-balances.

The large number of detailed pollution control laws makes rational priority setting difficult and discourages pollution prevention. The overlap and numerous inconsistencies among these laws are additional impediments to efficient and effective implementation. While costs of pollution reduction cannot be considered in setting national ambient air quality standards, for example, they are considered implicitly in setting technology-based water standards; explicit benefit-cost balancing is required by the toxics and pesticides laws and, since 1996, by the Safe Drinking Water Act.

The major problems with environmental legislation are their rigidity and lack of coherence. The laws are complex, unrelated to each other, and lacking in any unified vision of environmental problems or EPA's mission. This is not surprising given that the system of congressional committees and subcommittees dealing with environmental regulation is complex, the committees do not relate to each other, and there is no coherence to the their approach. EPA reports to literally dozens of congressional committees and subcommittees.

Most of the laws involve the establishment of ambient and source discharge standards, enforcement through inspections, and "incentives" to comply based on the threat of monetary penalties and sometimes jail terms. In practice, many of the laws dictate how the standards will be met—for example, what type of control mechanism must be installed to meet the standard. Some people refer to this as the "command and control" approach to compliance.

Programs designed to rely more on economic incentives are considered by economists and others to be more efficient than most approaches; but inertia, political constraints, and implementation questions have combined to make economic approaches a rarity. The outstanding example of an economic or "market" approach is the system of marketable allowances for emissions of sulfur dioxide from electric power plants contained in the 1990 amendments to the Clean Air Act. The system saved utilities and their rate payers an estimated $240 million dollars in 1995 without any sacrifice in environmental goals, and will save more than a billion dollars a year in the long run. The emissions trading system is, however, closely tied to the overall regulatory provisions of the Clean Air Act. Most kinds of environmental market approaches depend on the existence of a nonmarket regulatory framework.

The plethora of statutes, their detailed requirements, the traditional right of regulated industries to challenge the requirements, and the power that many of the laws give to citizens to initiate court action to enforce legal provisions all result in the courts' playing a significant role in the regulatory system. A majority of EPA's major rules are challenged in court. While the courts uphold most of the rules that are challenged, it is not unusual for a court to send a regulation back to EPA for revision. Such rules may subsequently be reissued, but the court's rejection is sometimes so strong or sweeping that for all practical purposes the rule is dead.

The opportunity to comment on proposed rules, coupled with the role of the courts, guarantees that most of the groups interested in pollution control regulations have some voice in what the rules will say. The right to appeal also serves to check arbitrary actions by EPA. However, because judges generally are not well versed in the science, economics, or administration of pollution control, their decisions may themselves be somewhat arbitrary.

Administrative Decisionmaking

At the federal level, administrative decisionmaking for controlling pollution is centered in EPA. The agency was created in 1970 by combining a number of programs that had been located in several different cabinet departments and agencies. The nature of its creation, the fragmented laws that give the agency its authority, and the fact that half its employees are located in ten regional offices make EPA something less than a solidly unified organization.

The centers of power within the agency are the three offices oriented to environmental media—air, water, and solid waste. The EPA administrator has more authority over the constituent parts than many federal agency heads, but the authority is limited by the fragmented statutes and, in some administrations, by the reluctance of the administrator to do battle with the media-oriented offices.

The culture of the agency is primarily legal in nature. Every EPA administrator except one (Lee Thomas) has been a lawyer. Emphasis on legal training is not surprising given the complexity of the laws, the importance of litigation, and the reliance on statutory provisions to accomplish the agency's goals. The legalistic orientation also is reinforced by the frequent interaction between all senior agency officials (both political appointees and careerists) and Congress. Congressional staff oversee every agency move of any consequence and do not hesitate to make their views known. Because Congress is even more dominated by lawyers than EPA, the mingling of executive and legislative branches reinforces the legalistic culture within the agency.

Although consideration of economics and science is important in most agency decisions, they receive less deep or serious attention than legal factors. For most of the agency's history, presidential executive orders have required the agency to conduct economic analyses of its major decisions. The quality of these analyses has varied widely; more importantly, the economic analyses often are not performed until after a decision has been reached. In the past, EPA's policy office oversaw the economic analysis process, a practice that served to prevent the program offices from using economic analyses to justify their previously taken decisions. Under current Administrator Carol Browner, the programs have been charged with performing their own economics.

Science in EPA takes a variety of forms, the most visible and probably most important being risk assessment. The results of risk assessments play an important part in many EPA decisions, and the agency recently reorganized its research laboratories to reflect different aspects of the risk assessment process. Scientific work in the agency is conducted in both the Office of Research and Development and the program offices, through both in-house and extramural researchers. The Office of Research and Development typically has not had much influence in internal agency deliberations, so the fate of scientific information is not unlike the fate of most economic information: important within the agency, but less important than other considerations, and with only weak checks to prevent program directors from distorting information to support decisions taken on other grounds.

There are mixed views about the overall quality of management at EPA. By almost any measure, the agency has a heavier workload than most other federal agencies, and this load often reduces the quality of management. However, the picture held by some members of Congress and some White House staff of an organization that acts impulsively and irrationally, ignoring both other agencies and common sense, does not have much resemblance to reality. There are no good measures of management competence, but it is likely that over its lifespan EPA has been managed at least as well as most other agencies.

EPA does suffer certain chronic administrative problems, but they are not unique to the agency. For instance, it often ignores the interests and capabilities of other agencies. It is largely devoid of any capability or desire to evaluate the programs it administers. Its internal information systems for allocating workload and measuring accomplishments are weak or nonexistent. Congress could help to remedy these problems, but more often than not it aggravates them.

Many other federal agencies play essential roles in the U.S. pollution control system. Arguably the most serious pollution problems in the nation are at sites owned by the Department of Energy. Nonpoint sources, the major contributer to water pollution, cannot be addressed without the cooperation of the Department of Agriculture. The Army Corps of Engineers is the primary regulator of the nation's wetlands. The Department of Health and Human Services, the National Aeronautics and Space Administration, the Department of the Interior, and the Department of Commerce all conduct more environmental research than EPA. The Council on Environmental Quality, the Office of Management and Budget, and the Office of Science and Technology Policy—all part of the Executive Office of the President—play important policy and coordinating roles in the pollution control regulatory system.

Because the major implementation burden for federal environmental statutes rests with the states, the most important agencies outside EPA are the state pollution control agencies.

The Federal-State Division of Labor

Although the federal government's involvement in pollution control dates back at least to 1910, state and local governments have always played the major role in implementing pollution control programs. The federal legislation of the early 1970s shifted much of the responsibility for setting standards and some enforcement responsibility from the states to the federal government, but the pollution control system is, like most federal programs, a complex mixture of responsibilities shared by several levels of government.

EPA issues national standards for air and water pollution and for other problems such as the design of solid waste landfills. However, implementation of the standards is based on permits issued to the sources of pollution, and generally the state agencies negotiate and issue the permits. In some programs EPA has authority to review the permits; in other programs it does not. Enforcement is conducted mostly by state and local agencies, but EPA and the Department of Justice also have enforcement powers. Most research is performed by federal agencies, but most monitoring of pollution levels is done by state and local agencies.

The three most important pollution control laws—the Clean Air Act, the Clean Water Act, and the Resource Conservation and Recovery Act—authorize EPA to formally delegate administration of important program responsibilities to a state if the state can demonstrate its ability to administer the program. Most state programs have received EPA approval to carry out these delegated responsibilities. As of 1995, thirty-eight states had authority to issue water permits, all but five states controlled hazardous waste under RCRA, and all but one state certified pesticides under the federal statute. Air delegations are currently being sorted out as EPA issues regulations to implement the 1990 Clean Air Act amendments.

The 1990s have witnessed several important changes in the federal-state relationship. The support that state pollution control agencies traditionally have given to EPA has turned to opposition with the increasing competence of state agencies, decreases in funding from the federal government, political currents favoring decentralization, and actions by EPA and Congress that impose expensive obligations on states and localities without providing funding to meet the obligations ("unfunded mandates"). The erosion of state support has significantly weakened the position of EPA and the national environmental organizations.

EPA has responded by instituting a "National Environmental Performance Partnership System." This program sets performance goals that states commit to achieving in exchange for much-reduced oversight by EPA. The basic concept of the proposal is sound, but success will require EPA regional offices to become cooperative partners with the states instead of adversarial overseers—not an easy change to accomplish.

All observers agree that the capabilities of state environmental agencies have improved significantly since 1970. The improvement in the professionalism and competence of state governments generally has been reinforced by the increased importance of the environment as an issue. Whether improved state capability argues for further decentralization of the already decentralized pollution control system is unclear.

There is general political sentiment that centralizing functions in the federal government is per se not desirable. Many knowledgeable observers think that centralization of pollution control functions in particular has gone too far and that more flexibility is needed to tailor standards to local conditions. Some feel that entire programs, such as drinking water or hazardous waste cleanup, should be returned to the states.

Those who defend federal powers point to competition for new industrial development among states and even with nations, contending that such competition will lead to pressure for increasingly more lenient environmental standards. Three other factors argue for a continuing strong federal presence:

- The relative political weakness of states and localities limits their ability to control the behavior of large multinational corporations.
- Global problems, such as climate change, are becoming a larger part of the environmental agenda.
- Many pollutants travel across state boundaries and thus cannot be controlled by the state that suffers the pollution effects.

This last problem, the original justification for federal enforcement authority, has never been adequately addressed by EPA. The agency is, however, taking some initiatives to deal with interstate pollution (in, for example, some of the provisions of the proposed 1997 ozone standard revision).

For the foreseeable future, neither EPA nor the states will have enough money to implement all of the legally required pollution control functions. Improvements in the federal-state relationship and in the division of labor between the two levels will probably occur, but they are unlikely to change significantly the overall effectiveness of the system.

EVALUATING THE REGULATORY SYSTEM

In the past decade the views of government officials, elected representatives, the media, and the public have increasingly diverged about whether the U.S. pollution control regulatory system is performing satisfactorily. Some people point to the significant reduction in most air pollutants, the cleanup of major rivers, and the tangible progress made in improving environmental quality. Others point to the inefficiency and intrusiveness of regulations and the lack of progress in dealing with nonpoint sources of pollution or with global climate change. The general public continues to strongly support pollution control, but at the same time often opposes the taxes necessary to pay for it and mistrusts the government officials who administer it.

Our project attempts to examine the full scope of the pollution control regulatory system and to evaluate it against explicit criteria. Not surprisingly, the conclusions that emerge provide a mixed picture—this is why different people perceive the system so differently. By identifying what parts of the system work well and what parts do not, we try to provide a context for understanding the system and an agenda for change.

Reducing Pollution Levels

We began by looking at actual levels of pollution in the environment and how these levels have changed over time. Our task was not easy because, for almost every type of pollution, monitoring data are woefully poor. We can tell something about national air pollution trends, although the data are often sparse. It is impossible to draw any firm conclusions on a national or state basis about whether water quality is improving or getting worse. Most other areas are like water quality: data are not adequate to determine whether conditions are improving over time.

Overall, the environment generally is better in 1997 than it was in 1970. It is certainly better than it was in 1940 or 1900, although such comparisons cannot be based on hard data and also involve subjective comparisons (such as between the effects of horse manure and diesel fumes). The regulatory system has prevented both population growth and the huge expansion of the U.S. economy from exerting large negative environmental effects.

Data for particular environmental areas are described below.

Air Quality

EPA concentrates its attention on six major air pollutants, called criteria air pollutants. Four of the six have shown significant improvement, both historically and in recent years. Between 1986 and 1995, average airborne concentrations of lead dropped 78 percent, while those of carbon monoxide and sulfur dioxide dropped 37 percent. Concentrations of particulate matter also have improved, although not as much. The regulatory definition of particulate matter changed in 1987 to focus on smaller particles, so long-term trend data are not available. The two criteria pollutants that have not shown as much improvement—nitrogen dioxide and ozone—are related to urban smog. Table 1 summarizes the improvements in criteria air pollutants.

The general improvement in air quality still has left some areas of the nation experiencing days when air pollution exceeds the national ambient air quality standards (NAAQS). As Figure 1 shows, ozone, an indicator of smog, accounts for most failures to attain standards. It is important to note, however, that a county is considered not meeting the standard even if the standard is only violated

Table 1. Air Quality and Emissions Trends 1986–95.

	Air quality change (%)	Emissions change (%)
Carbon monoxide	−37	−16
Lead	−78	−32
Nitrogen dioxide	−14	−3 (nitrogen oxides)
Ozone	−6	−9 (VOCs)
PM-10*	−22	−17
Sulfur dioxide	−37	−18

*PM-10 changes are based on 1988–95 data

Source: U.S Environmental Protection Agency. 1996. *National Air Quality and Emissions Trends Report, 1995.* Office of Air Quality Planning and Standards Research. EPA 454/R-96-005. Research Triangle Park, North Carolina; p. 1.

once or twice a year at only one of several possible monitors in the area; the figure in this sense may exaggerate the degree of the problem.

Figure 1. Population in Counties Not Meeting Ambient Air Quality Standards, 1993.

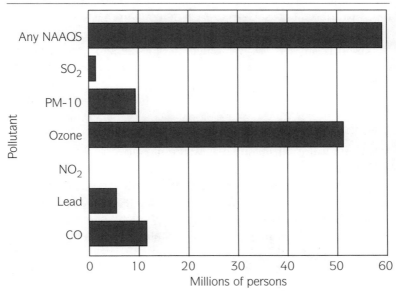

Source: U.S. Environmental Protection Agency. 1994. *National Air Quality and Emissions Trends Report, 1993.* Office of Air Quality Planning and Standards Research. EPA 454/R-94-026; Figure 5-9, p. 96.

Water Quality

A coherent picture of trends in U.S. water quality cannot be developed from existing data. Overall, it appears likely that the quality of the nation's rivers and lakes has improved dramatically in a few places but improved more modestly or stayed the same in most others. Water quality in estuaries and coastal waters probably has declined.

EPA's 1994 water quality report (EPA issues such reports biennially, based on state data) concludes that 43 percent of assessed rivers and 50 percent of assessed lakes fail to meet Clean Water Act goals that they be fishable and swimmable. This statement may exaggerate the extent of the water pollution problem because fewer than 20 percent of the rivers and only half the U.S. lakes were assessed; probably the more polluted ones received assessment.

The National Stream Quality Accounting Network (NASQAN), run by the U.S. Geological Survey, provides the only consistent data on water quality trends. The network data, while quite reliable, do not deal with most of the pollution situations that EPA currently worries about, such as oxygen demand from effluents around urban areas. The data are summarized in Table 2, which shows the trends in pollutants for two different time periods (for instance, between 1974 and 1981 dissolved oxygen pollution became much better).

During the most recent period, of the six water pollutants measured by NASQAN, all pollutants showed significant improvement. However, the measures deal only with rivers, and generally do not assess water quality in areas where most people live.

Looking at the sources of water pollution, EPA can claim one notable victory and must acknowledge one notable gap. The victory is in treatment of household or municipal sewage. For most of its history, the agency's largest program in dollar terms provided grants to local governments (through state governments) to build municipal waste treatment plants. In a 1988 evaluation to Congress, EPA reported that the number of people in communities receiving secondary or advanced levels of treatment rose from 4 million in 1960 to 143.7 million in 1988. While some 25 percent of the U.S. population is not served at all by public treatment facilities and other facilities are poorly operated, the massive increase

Table 2. NASQAN Trends.

Pollutant	1974–81	1980–89
Dissolved oxygen	Much better	Better
Fecal coliform	Much better	Much better
Dissolved solids	Not available	Much better
Nitrate	Much worse	Slightly better
Suspended sediment	Slightly worse	Much better
Total phosphorus	Slightly better	Much better

Note: The National Standard Quality Accounting Network (NASQAN), compiled by the U.S. Geological Survey, provides the only national profile of pollutant concentration trends in some rivers and streams during the 1970s and 1980s. The phosphorus data are for 1982–89, due to poor data quality in 1980 and 1981.

Sources: Smith, Richard A., Richard B. Alexander, and M. Gordon Wollman. 1987. *Analysis and Interpretation of Water Quality Trends in Major U.S. Rivers, 1974–1981.* U.S. Geological Survey Water Supply Paper 2307; p. 4. Smith, Richard A., Richard B. Alexander and Kenneth J. Lanfear. 1991. *Stream Water Quality in the Coterminous United States: Status and Trends of Selected Indicators During the 1980s.* U.S. Geological Survey Water Supply Paper 2400; pp.116–27.

in areas served by secondary treatment is definitely a major accomplishment.

The glaring gap in EPA's water quality program is with nonpoint sources, now the leading source of water pollution in most areas. Nonpoint sources include runoff from farms and city streets and also deposition of pollutants from the air into water bodies. While this type of pollution is more difficult to control than pollution coming from discrete outfall pipes at, say, factories or municipal treatment plants, a variety of measures could be taken. Neither Congress nor EPA has done much to address nonpoint sources, however, in part because of reluctance to take on the farm lobby. The $95 million per year EPA has given since 1990 to states to deal with nonpoint sources is less than 5 percent of the total funds it spends on water quality.

Drinking Water

The law governing federal efforts to assure the safety of drinking water was amended in 1996. Data on the effectiveness of the old program do not tell us much. The percentage of drinking water systems in compliance with federal standards hovered between 68 and 73 percent from 1986 to 1993. The number of violations rose and fell sharply, but the changes were due more to changed reporting requirements than to any changes in actual water conditions. Prevention of drinking water contamination is likely always to be more a function of local resources and efforts than of the federal program, although the new federal law contains a fund to assist localities in improving their drinking water facilities.

Municipal Solid Waste

The amount of garbage each of us generates has doubled since the 1960s. In large part, the amount reflects economic prosperity: people are buying, consuming, and discarding more things. The total amount of garbage generated has increased even more because there are more people producing it.

EPA encourages recycling by households, but local governments and private markets have the most influence on how much material is actually recycled. EPA regulates the safeguards that landfills and incinerators must employ, but the risks arising from these sources are not easy to detect, nor do good data exist on compliance of the facilities with the regulations. A significant accomplishment that can be credited to the Resources Conservation and Recovery Act (RCRA) is the virtual elimination of open burning of garbage, a widespread practice before the 1970s.

Hazardous Waste

Despite RCRA's broad tracking and reporting provisions for hazardous waste, few data exist to show whether the law is achieving its goals. What we do know is that whereas only a few years ago there was talk of a "crisis" in capacity for hazardous waste disposal sites, a surplus of disposal capacity now exists. What we do not know are the reasons for the surplus. It likely is due to more of the waste being treated on-site and to an increase in incinerators rather than to an actual reduction in the amount of waste generated. Whether this is a good trend for the environment is unclear. The RCRA program almost certainly has improved the methods used for handling hazardous waste. For example, land disposal of untreated hazardous waste has been greatly reduced.

The second largest EPA program (after water), as measured by direct program dollars spent, is the so-called Superfund program, designed to clean up sites contaminated by hazardous substances. The program has been a subject of controversy almost from its beginning. The major risks from hazardous waste sites have probably been addressed through emergency removal actions; 3,042 such actions had been undertaken as of 1995. Permanent cleanup of Superfund sites has proceeded more slowly, in part because negotiating the remedies to be applied and who should pay for them can be a slow, rancorous process and in part because the actual cleanup, especially of contaminated groundwater, can take many years. As of June 1996, construction work had been completed at 410 of the 1,227 sites on the Superfund National Priorities List.

Toxics

More than 2,000 new chemicals are reviewed each year by EPA through a process that is generally credited with significantly reducing the toxicity of new chemicals manufactured; unfortunately, no data exist to support the view. Although the manufacturer of a proposed new chemical is required to submit the results of any tests conducted on the chemical, only about 6 percent of the new chemical notifications sent to EPA contain any test data.

Seventy thousand chemicals are currently manufactured or processed in the United States. EPA has banned five substances under the Toxic Substances Control Act. The agency collects emission data on 343 chemicals (expanded to approximately 650 in 1995) under the Toxics Release Inventory (TRI), but because the TRI data cover only some emission sources the inventory underestimates emissions. Total TRI releases to the environment plus transfers (that is, wastes moved offsite) have increased from roughly 4.5 billion pounds to 5.5 billion pounds between 1988 (when the TRI process began) and 1994 (the latest year for which data are available). Total releases have declined by roughly 44 percent, but total transfers, including transfers to recycling and energy recovery, have tripled since 1988 (see Figure 2).

In 1991, EPA targeted seventeen of the most toxic substances on the TRI list for reduction under a voluntary program called

Figure 2. TRI Releases and Transfers: 1988, 1992–94.

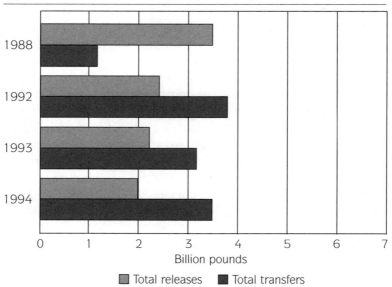

Note: Data used for year-to-year comparison differ from 1994 reporting
year totals; 1988 transfer data are not comparable with other reporting
years.

Source: U.S. Environmental Protection Agency. 1996. *1994 Toxics Release
Inventory: Public Data Release;* Table 3-3, p. 171.

"33/50." The program achieved its goal of a 50 percent reduction
in emissions for these chemicals. Long-term production figures for
the chemicals show a mixed picture, however. Between 1970 and
1990, production declined for three of the chemicals, increased for
five, and was not significantly changed for six. Data were not available for three of the chemicals. The government has stopped col-
lecting data on chemical production because of budget reductions.

The fragmentary data on the level of toxics found in humans
and wildlife indicate that significant progress has been made in
addressing some persistent toxics. The average level of lead found
in the blood of humans has declined 78 percent from 1976–80 to
1988–91. The level of PCBs, long-lived chlorinated compounds that
may cause cancer, has declined in Lake Michigan trout from a peak

of twenty-three in 1974 to less than three in 1990 (measured in micrograms per gram wet weight). Similar reductions have been achieved for other chlorinated compounds (such as DDT) and in other locations.

Pesticides

Pesticides, including fungicides, herbicides, and similar products, are a unique group of chemicals because they are intended to be toxic to some living things and they are deliberately introduced into the environment. Because of this combination they are closely regulated, and EPA reviews and approves every pesticide for every particular use.

More than half of pesticide use is on agricultural crops, so the volume of pesticides used tends to vary with the volume of agricultural production, the crop mix, and the technologies used in agriculture. Total pesticide use in agriculture increased from 232 million pounds in 1964 to 573 million pounds in 1992. However, the peak year was 1982 (612 million pounds). The mix of pesticides used has changed significantly. Between 1964 and 1992, use of insecticides declined by 60 percent while use of herbicides increased by more than 600 percent. Total U.S. production of pesticides has remained at about 1.1 billion pounds for each of the past ten years.

As noted above, the level of persistent toxics, including pesticides, in humans and wildlife has declined significantly. Newer pesticide compounds are less persistent in the environment. They tend to be more acutely toxic than the older types of pesticides but are less likely to cause chronic problems such as birth defects. The revolution in genetic engineering is likely to have a major impact on the amount and types of pesticides used in the future, but it is too early to know the environmental implications of this change.

How much of any change in pollution levels is due to pollution control regulatory programs? We know that both weather (especially temperature and precipitation) and economic changes directly influence pollution levels apart from any pollution control efforts. This fact is illustrated, for example, by the figure for the number of people living in areas that are not meeting one or more of the national ambient air quality standards (see Figure 1, page 17). The figure fluctuates widely from year to year because it is driven primarily by nonattainment with the standard for ozone, whose atmospheric levels are heavily influenced by average summer temperature. It is not far-fetched to say that these nonattainment figures really measure how hot the last summer was. The ozone standard proposed by EPA in late 1996 would alleviate this problem somewhat by basing the standard more on statistical averaging.

In an effort to statistically disentangle these factors, we analyzed economic, meteorological, and air pollution control cost data for a twenty-year period in three metropolitan areas where the level and composition of economic activity substantially changed concurrently with the implementation of national air quality standards: Allegheny, Pennsylvania (which includes the city of Pittsburgh); Baltimore, Maryland; and Cuyahoga, Ohio (Cleveland). Air quality in these areas has improved dramatically over the past few decades, but the improvement is often attributed to the decline in heavy manufacturing, leaving the impression that mandated pollution control investments have had little impact. Previous studies have found little evidence of an association between pollution control investments and air quality.

The results of our analysis suggest that mandated pollution control investments often have had a significant effect on reducing air pollution levels. Because we found such an effect in areas where economic changes have been stark, it is reasonable to assume that significant regulatory effects also have occurred where changes in the level and composition of economic activity have been less dramatic. However, in the three areas we examined, the effects of economic changes, weather, and other factors were much greater than the effects of regulatory controls. Also, the failure of local factors (such as the level of local manufacturing activity and local pollution control investments) to account for a majority of the variation in local air quality underscores the importance of regional or national factors (both regulatory and nonregulatory) in determining local air quality.

Targeting the Most Important Problems

Regardless of how effectively EPA reduces the levels of pollutants on which it is focused, if it is focusing on the wrong targets it may not be doing much to address protection of human health and the natural environment—the real goals of the pollution control regulatory system. Investigating this question takes us into the realm of priority-setting and comparative risk.

Although we talk about EPA's priorities, the priorities are really set primarily by Congress. As we noted earlier, statutes drive the pollution control priorities, and EPA has limited flexibility to adjust legislatively established priorities. Furthermore, setting priorities is dependent on other factors in addition to consideration of the risk presented by a problem. The cost of control measures, their implementability, and how the costs and benefits are distributed are just some of the other dimensions that determine program priorities. However, the relationship between priorities and risk is a fundamental criterion that should be considered in evaluating the pollution control system.

The best indicator of EPA's priorities is the money spent on particular problems. Most analyses focus on EPA's budget expenditures, but this tells only part of the story: the real impact of EPA programs often depends on expenditures by state and local governments and the private sector. The private sector bears about 60 percent of total pollution control costs; local governments account for 20 to 25 percent of total costs; EPA and other federal agencies account for less than 10 percent. States supply the remainder.

EPA's budget priorities have remained fairly constant since the agency was created. Water pollution control has topped the list, followed by hazardous waste and air pollution. At this general level, private sector spending priorities are similar to EPA's, although the private sector spends more on air than hazardous waste. At the level of specific programs, greater disparities between public and private spending priorities appear. For example, the Superfund program in fiscal year 1994 accounted for 26 percent of EPA's budget but only 3 percent of projected total U.S. pollution control expenditures. Few reliable data exist on the actual cost of Superfund site cleanup. Figure 3 shows the cost of complying with EPA regulations both by environmental medium and by who bears the costs.

The first comparative assessment of the risks controlled by EPA programs was performed in 1987 by a team of senior EPA analysts. Since then, the EPA Science Advisory Board has conducted

Figure 3. Total Pollution Control Costs by Medium and Funding Source, 1986.

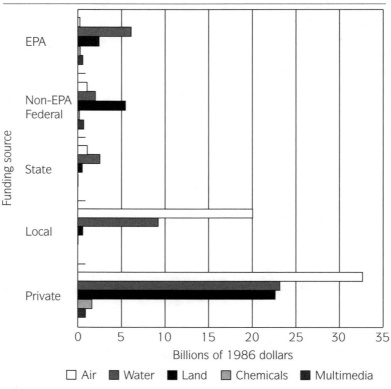

Source: U.S. Environmental Protection Agency. 1990. *Environmental Investments: The Cost of a Clean Environment.* Report to the Administrator; Table 8–12A, p. 8-51.

two extensive reviews of the 1987 report, each of the ten EPA regions has performed a comparative risk assessment, and twenty-five states and several localities have compiled risk rankings. The methods used by the states varied widely, and the rankings were done at different times. On the whole, however, the results were quite similar. Figures 4 and 5 (see pages 28 and 29) show the health and ecological risk rankings from the EPA regions.

Comparing these rankings with expenditures by either EPA or society at large reveals a huge gap between risk priorities and spending priorities. Indoor radon, which tops the health concerns of EPA regional officials, is regulated not at all by EPA and only minimally by the states. The EPA radon program is 0.07 percent of the EPA budget. Indoor air pollution, the second-ranked health problem, similarly is not for the most part regulated by EPA or the states. Local building codes address some aspects of the indoor air problem, such as rates of air exchange. EPA has an indoor air program—it is 0.24 percent of the agency's budget.

The story on ecological risks is not much better. Nonpoint source pollution is rated the most serious ecological risk; to the extent that it is regulated at all, it is regulated by the states. The EPA nonpoint program has in recent years accounted for 1.7 percent of its total budget. Stratospheric ozone depletion, the number two ecological risk, is being regulated; in fact, EPA's regulation of aerosol propellants was the pioneering initial move in addressing the stratospheric ozone problem. Pesticides, which rank third on both the health and ecological risk rankings, are regulated by both EPA and the states. The EPA pesticides program, however, has received only 2.4 percent of the EPA budget.

All questions of comparative risk are plagued by the inadequacy of information about the nature and severity of environmental problems. There are not enough toxicity data on most chemicals to know whether they cause adverse effects. There are not enough monitoring data to know to which pollutants people are exposed. We do not understand many fundamental aspects of the earth's ecology—we do not understand the role of clouds in the earth's temperature balance or what makes flowers bloom. Knowledge about how pollutants travel from one part of the environment to another is woefully inadequate. These are problems both of fundamental scientific knowledge and of inadequate data collection.

Another difficulty with comparative risk is the effect of past and existing control efforts. Drinking water, for example, would likely be the major environmental health threat were it not for existing water purification efforts. Nonpoint sources of stream pollution are so important now in part because of past success in controlling point sources. The results of comparative risk analyses therefore should be used primarily as guidance for allocating future efforts, not as a justification for dismantling current efforts.

Figure 4. Human Health Risks Ranked As High by EPA Regions.

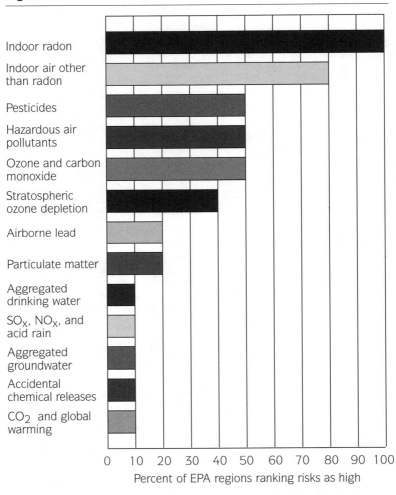

Indoor radon

Indoor air other than radon

Pesticides

Hazardous air pollutants

Ozone and carbon monoxide

Stratospheric ozone depletion

Airborne lead

Particulate matter

Aggregated drinking water

SO_x, NO_x, and acid rain

Aggregated groundwater

Accidental chemical releases

CO_2 and global warming

0 10 20 30 40 50 60 70 80 90 100

Percent of EPA regions ranking risks as high

Source: U.S. Environmental Protection Agency. 1993. *State and Local Comparative Risk: Project Rankings Analysis.* Office of Policy, Planning, and Evaluation, p. 20.

As we have noted, comparative risk is not the only relevant factor in determining expenditure levels or priorities. The potential effectiveness of government action, the cost of dealing with a problem, and numerous other factors are important to consider. However, the

Figure 5. Ecological Health Risks Ranked As High by EPA Regions.

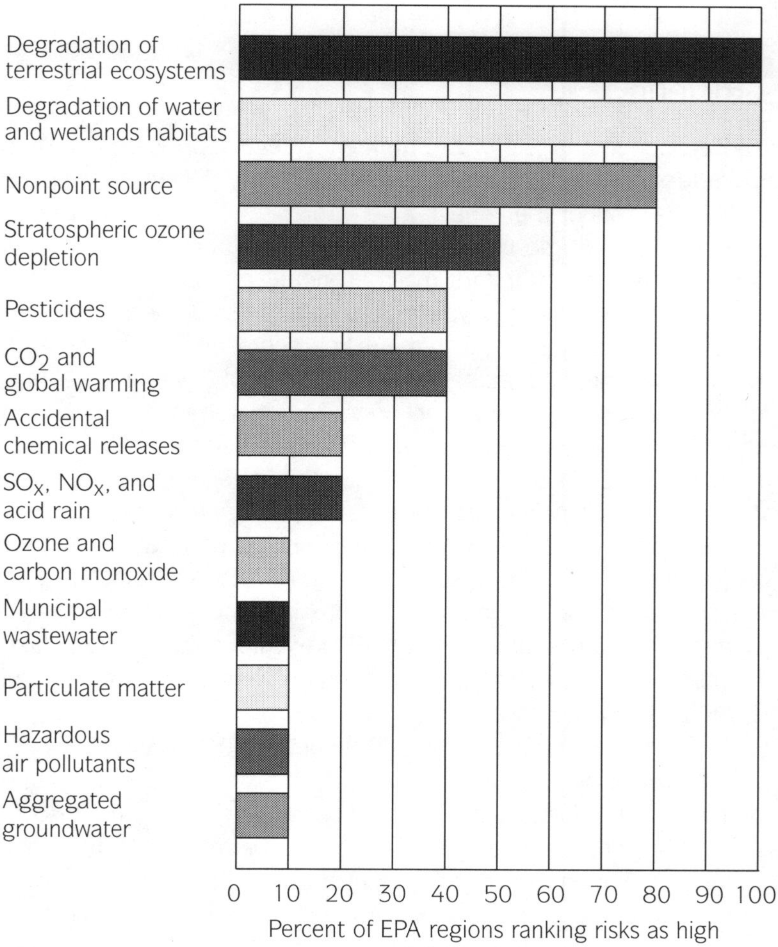

Degradation of terrestrial ecosystems
Degradation of water and wetlands habitats
Nonpoint source
Stratospheric ozone depletion
Pesticides
CO_2 and global warming
Accidental chemical releases
SO_x, NO_x, and acid rain
Ozone and carbon monoxide
Municipal wastewater
Particulate matter
Hazardous air pollutants
Aggregated groundwater

0   10   20   30   40   50   60   70   80   90   100
Percent of EPA regions ranking risks as high

Source: U.S. Environmental Protection Agency. 1993. *State and Local Comparative Risk: Project Rankings Analysis.* Office of Policy, Planning, and Evaluation, p. 21.

conclusion seems inescapable that the pollution control regulatory system is not addressing some very important problems. It is focusing almost exclusively on outdoor air pollution when a large part of the health risk comes from indoor air pollution. It is focusing on point sources of water pollution when the major problem today is nonpoint sources. The current priorities are driven by the pollution control statutes. Until these statutes are changed, the priorities will be askew.

Efficiency

The economic analysis of environmental regulation is complicated by at least two major problems. First, it is very difficult to evaluate either the costs or the benefits of environmental programs and regulations. Second, most of the benefits of environmental regulations are excluded from standard macroeconomic indicators: cleaner air, for example, is not a gain when calculating gross domestic product.

The difficulty of calculating environmental benefits is due both to the uncertainty regarding the actual effects of environmental regulation—for instance, fewer sick days, less materials damage—and to the lack of any widely agreed-upon way to attach a dollar value to many of the effects. For example, a major benefit of the Clean Air Act is a reduction in mortality from respiratory disease. But how many deaths are avoided? What value should be attached to prolonged life? Should the age of those whose death is forestalled be considered—since everyone dies eventually, should we really be calculating and attempting to value life-years saved? These and numerous other questions lacking good answers must be addressed in the benefits calculation.

Estimating regulatory costs is somewhat easier, but is still rife with potential pitfalls and uncertainties. The true costs of a regulation include not just the impacts on the regulated entity but also its ripple effects throughout the economy. If paper manufacturers have to install pollution control equipment, they probably will pass some of the cost on to customers, which means that the cost to all businesses that use paper will increase. Calculating these costs is not simple. (An analogous problem exists with benefit estimation: environmental regulations have benefits, such as encouraging the manufacture of pollution control equipment, which are usually not counted in the standard analyses.)

Environmental regulation can also give rise to costs that cannot be measured by observing pollution control spending—economists call these opportunity costs. Examples would be the time that people spend waiting for vehicle inspections or the new plants that might not be built because of strict standards on new sources of pollution. Transaction costs also often are significant—it can take a long time to get a permit or site a facility, and, in this context, time is money. An added difficulty in cost estimation is that the private sector often develops ingenious ways of reducing the costs of regulation when it is permitted to do so; as a result, the early cost estimates are too high—they fail to take into account the inventiveness and flexibility of the private sector.

The several attempts over the past fifteen years to estimate the aggregate benefits and costs of the Clean Air Act and Clean Water Act suffer from significant limitations, but their findings are consistent.

Taken as a whole, the benefits of the Clean Air Act seem clearly to outweigh the costs. Three separate studies using different methods and different data sources conclude that aggregate benefits exceed aggregate costs. One study that looked just at EPA rules promulgated between 1990 and 1995 concluded that the benefits of the Clean Air Act rules in this period exceeded the costs by about $70 billion (in 1994 dollars).

The story seems to be quite different for the Clean Water Act. There also have been three attempts to estimate its aggregate benefits and costs, and again all three came to the same conclusion. However, in contrast with the Clean Air Act, the conclusion was that the costs of the water act exceeded its benefits. There are great uncertainties with all of the benefit and cost estimates, and a number of tangible benefits which are not included in the benefit-cost analyses. Nevertheless, the contrast with the studies of clean air costs and benefits is striking. Probably the major difference between the air and water acts in this context is that the benefits of the air act are primarily improved human health, whereas the benefits of cleaner water are primarily ecological and recreational. Society tends to place a higher value on health benefits.

A number of studies have taken a different approach to judging the efficiency of the pollution control regulatory system. Rather than looking at entire programs, they have looked at individual regulations. The results indicate significant inefficiencies in the system.

One of the most comprehensive of these studies examined abatement costs for the same pollutant across different industries.

A significant disparity in the marginal cost of abating pollution (that is, the cost of abating the next increment of pollution) among plants, firms, or industries indicates that maximum environmental protection is not being obtained for the amount of money being spent. The study examined air pollution abatement costs between 1979 and 1985 by economic sectors. The variances among sectors were often quite wide. For example, the marginal cost of abating a ton of lead was $46,612 in the food sector but only $11 in the non-metal products sector; abating a ton of hydrocarbons cost the beverages industry $11,918 at the margin, but cost the pulp and paper industry only $20. The system could have reduced hydrocarbons much more efficiently by getting more reductions from pulp and paper plants and fewer from beverage makers. There are many such examples.

Several studies have looked at the cost-effectiveness of health regulations by measuring the dollars spent per life saved. These often are very partial studies because lives saved is usually based only on the results of risk assessments for cancer, thus ignoring other health benefits and all benefits other than health. Also, the numbers used in these studies are hugely uncertain. Nevertheless, the disparities among regulations are large enough to indicate that there is much room to improve cost-effectiveness. These studies also indicate that to the extent that the goal of environmental regulation is to save lives it often is more efficient to invest in nonenvironmental programs (such as encouraging motorists to wear seat belts) or in programs for early detection of serious diseases.

Finally, a number of studies have compared the costs of actual regulations with what the costs might have been for other environmental policy approaches, such as economic incentives. These comparisons suffer from all the problems of comparing an actual approach with a theoretical one. Nevertheless, they consistently find that approaches other than command-and-control regulation would be more efficient. A recent review of twenty-seven such comparisons found that command-and-control regulations cost one-and-a-half to more than twenty times as much as incentive-based approaches. The few examples of actual use of economic mechanisms, such as the sulfur dioxide emissions trading provisions of the 1990 Clean Air Act, tend to confirm the theoretical studies.

In sum, ample evidence indicates that inefficiency is a problem for the pollution control regulatory system. The costs of some regulations may exceed their benefits by any definition of costs and benefits. While benefit-cost analysis is too crude a tool to use as the sole criterion of regulatory decisionmaking, when it indicates a large excess of costs over benefits, a responsible decisionmaker needs to have some good reason for proceeding with the regulation. Efficiency is also a problem when approaches to compliance other than the current form of regulation may achieve the same goals at much lower cost.

Social Values

Some of the most vehement objections to environmental regulation have had nothing to do with its lack of efficiency or effectiveness. These objections have instead been based on the belief that regulations conflict with important social values—values such as privacy, the right to participate in decisions that affect one's life, due process, protection of private property, and nondiscrimination.

Environmental protection itself has become a basic social value. This is shown by numerous public opinion polls as well as by the public outcry that occurs whenever environmental values appear to be seriously threatened. The criticisms of pollution control efforts based on social values need to be placed in this context: the American people expect pollution to be controlled and they expect their government to do it.

Evaluating programs based on their adherence to or violation of social values is difficult. With no objective measures for most values, judgments tend to be both strongly held and very subjective. There are intelligent, responsible people who think that the pollution control regulatory system is a fundamental threat to American civil liberties. This notion strikes us as absurd, but it is difficult to muster logic or data to "prove" the case one way or another.

Public Involvement

Public participation, the idea that people should have a voice in decisions that affect them, is probably the most salient social value in the regulatory system. The Administrative Procedures Act requires regulatory agencies to allow the public to comment on proposed rules and requires the agency to respond to comments. Many public participation mechanisms are employed by EPA, and both general laws (such as the Freedom of Information Act) and pollution control laws delineate methods for allowing public views to influence government decisionmaking.

Although public participation is generally considered an unalloyed good, remarkably little serious analysis exists about what forms of participation work best, how the public can be most effective in its participation, and how participation mechanisms actually affect agency decisions. Our consideration of participation was handicapped by the lack of such analysis.

Environmental agencies at both the state and federal levels often have been in the forefront of experimenting with new forms of public participation. For example, EPA was one of the first federal agencies to use regulatory negotiation—a formal convening of interested parties—as a way of formulating regulations. More recently, the agency has turned to other forms of stakeholder participation in its Common Sense Initiative, XL projects, and place-based initiatives. Measuring the degree, much less the effectiveness, of participation is difficult, but our overall impression is that EPA is among the more "open" federal agencies.

Unfortunately, EPA's openness does not seem to have earned it much trust or credibility. For the past twenty-five years the American people have been increasingly mistrustful of all institutions, and especially of their government. The corrosive effect upon public actions is very clear with respect to environmental functions. People object to siting facilities and they ignore warnings of health threats; decisions about allowing or not allowing products to be sold are immensely difficult to implement because of people's mistrust of governmental authorities. Large portions of the public do not believe what government officials tell them. The problem is not unique to environmental officials, and there are no simple remedies.

Any long-run solution to the trust problem requires government officials to be completely open and honest in their dealings with the public. State environmental agencies vary widely in their degree of openness. The federal EPA, as we have noted, is relatively open, although it can be faulted for peddling a false certainty about some of its conclusions (for example, when giving only a single number as the conclusion of its risk assessments instead of a range). Congress often deceives the public by conveying the notion that environmental regulations can provide absolute safety: scientific findings indicate that for many, perhaps most, pollutants safety is a matter of degree and absolute safety is unobtainable. Both Congress and the executive branch shrink from conveying to the public the difficult tradeoffs that environmental decisions often entail, and the public encourages such deception by wanting absolute safety at no economic cost in the same way it wants more government services and lower taxes. A better-educated public would reduce the temptation of government officials to practice such deception.

Intrusiveness and Paperwork

Protection from intrusiveness is another social value that the pollution control regulatory system is often accused of violating. We define intrusiveness as the characteristic of an action or requirement that results in an unwelcome or uninvited imposition on the public, including direct cost, lost time, restricted options, invasions of privacy, and inconvenience. For this project, we developed a crude and quite subjective "intrusiveness index" to rank the major environmental programs. We found a fairly wide spectrum, with the National Environmental Policy Act being the least intrusive and air pollution control programs the most intrusive.

Like most social values, protection from intrusiveness is embedded in a number of social values and cannot be considered in isolation. The general perception of recycling provides a good example. By our ranking, recycling programs are at least as intrusive as most other pollution control programs; yet they do not evoke the same negative reactions, perhaps because recycling is perceived as directly doing something good for the environment.

An opposite example is recordkeeping requirements, which many business people consider among the worst evils of the pollution control system. Probably this is because a lot of recordkeeping is viewed as redundant and unnecessary rather than because it is more intrusive than other requirements. The Paperwork Reduction Act requires most federal agencies to submit annually to the Office of Management and Budget (OMB) estimates of the time required

by businesses and state and local governments to complete reports
required by the agency. This information provides one measure of
intrusiveness. In 1995, all federal agencies combined imposed a
total of 6.9 billion hours of reporting. The Treasury Department
imposed 5.3 billion of these hours, mostly through tax filing requirements imposed by the Internal Revenue Service. EPA imposed 104
million hours, more than the departments of Transportation and
Education but less than several other federal agencies. The
Department of Labor required 266 million hours, the Securities and
Exchange Commission 191 million hours, and the Federal Trade
Commission 146 million hours. Between 1988 and 1995, EPA's
paperwork requirements grew by about 50 percent, but the requirements imposed by six other agencies grew at an even higher rate,
some more than doubling in size.

OMB does not report the paperwork requirements by individual regulatory program, but the Chemical Manufacturers
Association analyzed the 1994 EPA data by program. It found a
wide range. The water pollution control program imposed three
times the burden of any other EPA program, requiring 29 million
hours of paperwork. Next were the air program with 9 million and
the Resource Conservation and Recovery Act program with 7 million. The Safe Drinking Water Act program was lowest, imposing
only 11,000 hours of paperwork.

Environmental Justice

Environmental justice is a recent and quite sensitive social value
affecting pollution control. Advocates for environmental justice have
demanded increased regulatory oversight and more stringent environmental protection measures for minority and low-income communities and workers. Shortly after taking office, EPA Administrator
Carol Browner listed environmental justice as one of her top priorities, and in 1994 President Clinton issued Executive Order 12898,
Federal Actions to Address Environmental Justice in Minority
Populations and Low-Income Populations.

EPA program offices have developed training programs for
both EPA and state regulatory officials that examine how environmental justice concerns can be integrated into routine agency functions. In 1994 the agency began a small grants program to assist

grassroots organizations and tribal governments in outreach work and in employing technical experts to analyze and interpret environmental data. However, it is not clear how much EPA can actually do to address perceived environmental inequities. Many of them spring from forces that are not within EPA's control—local land use practices, market dynamics, and state and local agency enforcement activities.

How environmental justice should be defined is not clear; neither is it clear, under most definitions, how serious the environmental justice problem is. Just the methodological problems of examining the issue are immense. For example, conflicting conclusions about equity emerge depending on the geographical unit used in an analysis. Measuring the distribution of environmental burdens by census tract can give quite different conclusions than measuring it by census block, zip code, or county.

A final complicating point is that environmental justice is not cost-free. At least in theory, to the extent that priority is given to addressing the environmental needs of minorities and the poor, it may not be given to actions that may protect a larger number of people. The overarching strategy of environmental protection in this country has been based not on a standard of justice that assumes government regulation should be directed to improving the conditions of some particular members of society but on utilitarian principles—the greatest good for the greatest number of people. These utilitarian principles are incorporated into environmental policies through such tools as benefit-cost analysis and, more recently, comparative risk. The potential conflict between environmental justice and utilitarianism will not be easy to reconcile.

Comparison with Other Countries

For most of the twenty-five years that environment has been on the agenda of all nations, the United States has been a leader. The United States was one of the first countries to launch a major national pollution control effort; it was a major force behind the 1972 Stockholm conference that put environment on the international agenda; and American environmental standards and institutions have been copied throughout the world.

Comparisons between the United States and other countries reveal a more mixed picture. A detailed comparison of pollution control standards is impossible because much of the impact of standards depends on technical details like measurement times and methods. Obviously, the degree of compliance with the standards also is critical. Most U.S. ambient and source discharge standards probably are at least as stringent as those in other countries, and the U.S. record of compliance is probably better. On some important specific requirements, such as removal of lead from gasoline, we know that the United States is ahead of most other nations.

Actual levels of environmental quality are difficult to compare for the same reasons that standards are hard to compare. The technical details of monitoring can make a large difference, and in most countries (including the United States) ambient monitoring data are sketchy. To the extent that data exist and can be roughly compared, and based on subjective impressions, ambient air and water quality in the United States compares favorably with that in most other countries.

The data on emissions of air pollutants tell a different story. The United States accounts for a huge proportion of the world's emissions of such major pollutants as carbon monoxide, nitrogen oxides, sulfur dioxides, and carbon dioxide. It emits per capita more than two times as much carbon dioxide as Japan, more than three times as much nitrogen oxides as France, and more than eight times as much nitrogen oxides as Japan.

Although some of the disproportionate air pollution emissions in the United States are attributable to the size of the U.S. population, geographic area, and economy, these factors do not explain a large part of the difference between the United States and other nations. When the emissions data are controlled for such factors, a significant difference still exists between the United States and all other countries except Canada (see Figures 6, 7, and 8). Physical factors like climate may account for some of the difference, but it is hard to avoid the conclusion that the U.S. lifestyle is a significant part of the explanation. We live in bigger houses, drive more, and generally use more energy than other countries. While there are advantages to all this, it comes at an environmental cost.

The environmental impact of the U.S. lifestyle is not limited to air emissions. Per capita municipal waste generation in the United

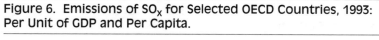

Figure 6. Emissions of SO_x for Selected OECD Countries, 1993: Per Unit of GDP and Per Capita.

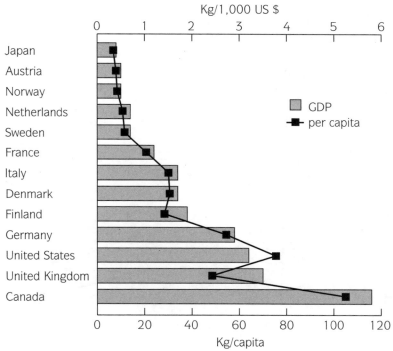

Note: GDP figures are given in terms of 1991 prices and purchasing power parities.

Source: Organisation for Economic Co-Operation and Development (OECD). 1996. *Environmental Performance Review, United States.* Paris; pp.254–55.

States is twice the level of such developed nations as Germany, Italy, and England.

The United States may be falling behind other countries with respect to institutional practices. In September 1996, after several years of debate, the European Commission approved a directive that requires all countries in the European Union to adopt an integrated approach to pollution control, an approach that cuts across the lines of media (air, water, land) on which the U.S. system is based. In four or five years, when the E.U. directive is fully imple-

Figure 7. Emissions of NO_x for Selected OECD Countries, 1993: Per Unit of GDP and Per Capita.

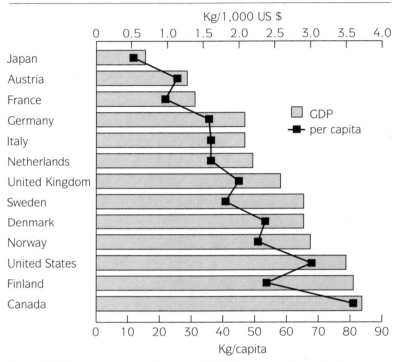

Note: GDP figures are given in terms of 1991 prices and purchasing power parities.

Source: Organisation for Economic Co-Operation and Development (OECD). 1996. *Environmental Performance Review, United States.* Paris; pp.254–55.

mented, the United States will be one of the few developed nations that has not adopted an integrated approach. It will be one of the few industrial countries issuing separate facility permits for air and water and not considering the tradeoffs among emissions to air, water, and land.

Market mechanisms also may be used more in countries other than the United States, although in Europe pollution taxes often are used more to raise revenues than to affect the behavior of polluters. The United States has pioneered emissions trading and other impor-

Figure 8. Emissions of CO$_2$ for Selected OECD Countries, 1993: Per Unit of GDP and Per Capita.

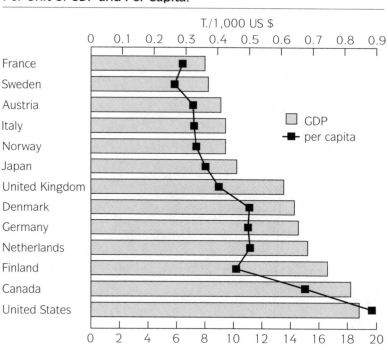

Note: GDP figures are given in terms of 1991 prices and purchasing power parities. CO$_2$ from energy use only; international marine bunkers are excluded.

Source: Organisation for Economic Co-Operation and Development (OECD). 1996. *Environmental Performance Review, United States.* Paris; pp.254–55.

tant economic approaches to environmental problems, but other nations, especially in Europe, may be getting more environmental mileage from applying market mechanisms. Taxes on gasoline are a good example, as shown in Figure 9.

There are other innovative environmental programs from which the United States could learn some useful lessons. In addition to integrating their approaches to pollution control, countries are experimenting with voluntary industry agreements on emission

Figure 9. Gasoline Prices and Tax Component (in dollars per gallon) in Selected OECD Countries, 1995.

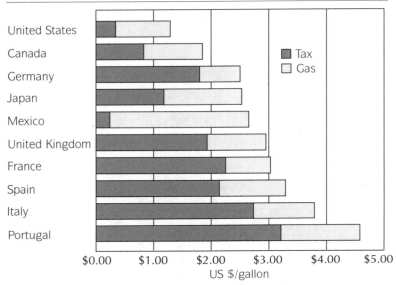

Source: RFF calculations based on *OECD Environmental Data Compendium 1995;* Table 9.5b, p. 230.

reductions (the Dutch covenants), recycling requirements (the German law that makes packaging disposal the responsibility of the manufacturer), and energy conservation. Of course, major cultural differences need to be taken into account—in Japan, for example, saving face is a major motivating factor. Programs based on this incentive would not work well in the United States.

International trade is becoming increasingly important in the U.S. economy, and in recent years the numerous connections between trade and the environment have come to be recognized. To date, the U.S. pollution control regulatory system has probably not been a significant drag on American trade abroad, although U.S. firms clearly would benefit from a more efficient regulatory system. Incorporating and reconciling environmental considerations into the international trade regime under the World Trade Organization is one of the major challenges facing the countries of the world.

Ability to Meet Future Problems

The problems addressed by the pollution control regulatory system are not static. Existing problems change for better or worse, and new problems are added to the agenda. In the background, a small but nagging chorus predicts that disaster is imminent and that most current programs resemble the proverbial rearranging of deck chairs on the Titanic. The dynamic nature of environmental problems makes the ability to anticipate and deal with future problems an important criterion for evaluating the system.

One way of looking into the future is to look at the underlying trends that affect pollution levels. These trends do not all go in the same direction—some predictable trends will result in worse environmental quality, while others will improve the environment. The steady increase in vehicle miles traveled makes it harder to achieve satisfactory air quality. However, the steady trend towards "dematerialization," using less material to perform the same function (for instance, the transistor supplanting the vacuum tube), makes it easier to deal with waste problems. Individual problems must be examined in the light of individual trends—broad generalizations are likely to be misleading.

A look at energy use and other trends affecting air pollution emissions suggests that the level of most major pollutants will continue to decline over the next fifteen years (see Figure 10). Two exceptions are carbon dioxide and particles. For both of these pollutants, emission levels are closely related to energy consumption. EPA predicts that by 2010 emissions of particles (PM-10) will be almost 60 percent higher than they were in 1990.

Water quality and water quantity are closely related because both affect water use. Because many people think that availability of water will be the dominant environmental problem of the twenty-first century, we looked at water supply projections for the United States. The picture is complicated by regional variations, lack of data, distinctions between withdrawal and consumption, in-stream versus non–in-stream uses, dry years versus wet years, and so forth. Several simplified conclusions are nevertheless possible. Projecting to 2040, only the Lower Colorado and Rio Grande river basins are likely to face major water shortages. Shortages appear-

Figure 10. Normalized U.S. Total Anthropogenic Emission Projections (1990 = 100).

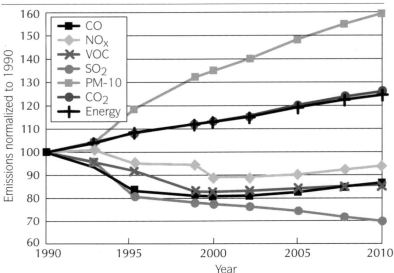

Source: U.S. Environmental Protection Agency. 1994. *EPA National Air Pollutant Emission Trends, 1900–1993;* U.S. Department of Energy/Energy Information Agency. 1994. *Annual Energy Outlook.*

ing elsewhere will be due more to institutional problems than to physical limitations. The overwhelmingly largest proportion of water use is for agricultural irrigation, and if mechanisms that facilitate transfer of water rights from agriculture to other uses can be encouraged, most of the shortages will disappear. This shift can occur without significant damage to American agriculture. The most severe problem is in the Lower Colorado basin: even if all agricultural uses of water were stopped there, a water supply deficit would still exist.

We foresee no startling changes or problems with the generation of municipal solid waste over the next couple of decades. Increases in population are likely to be offset by reduced waste disposal per person because of increased recycling. We also examined the problem of heavy metals, particularly chromium, mercury, and arsenic. Because heavy metals accumulate in the environment, they will almost by definition present an increasingly serious problem. We were frustrated in our attempts to estimate the rate of increase of metals in the environment because a large portion of the environmental load comes as a by-product of other processes, especially the burning of coal. For example, 25 percent of total chromium emissions in the United States comes from fossil fuel combustion. This relationship not only makes it difficult to estimate the magnitude of the problems but also indicates a possible weakness in the regulatory system: although a small amount of heavy metal emissions from fuel combustion is controlled by controlling particulate matter emissions, there are no direct controls on heavy metal emissions.

Completely new problems can appear on the environmental agenda. Nobody in 1970 thought that stratospheric ozone depletion was a problem. Five or ten years ago, not many people were worrying about endocrine disrupters. Identifying problems before they become major is inherently speculative, but it is important to try. EPA to its credit has in recent years recognized that future forecasting is an essential part of its responsibility. The purpose is not to predict the future but rather to examine the implications of current programs and to get a faster start on the next generation of problems.

Forecasting also highlights the degree to which pollution control is intertwined with policies in other areas such as energy, agriculture, and transportation. The U.S. government lacks ways to bridge across these different realms (with the notable exception of environmental impact statements). However, bridges need to be found, because the future of environmental quality will depend on what we do in these major economic sectors.

TOWARDS A BETTER SYSTEM

Given the scope and complexity of the pollution control system and the variety of criteria that we have used to evaluate it, we conclude not surprisingly that it has both strengths and weaknesses. Its greatest strength is its proven ability to reduce conventional pollutants generated by large point sources such as power plants and factories. It is a system that was developed to deal with the problems of the 1960s and 1970s, and it did a reasonably good job of addressing them.

In the course of dealing with those problems, the system has developed several other positive attributes. It opened the way to a variety of citizen efforts, ranging from recycling to regulatory negotiation and from citizen suits to monitoring the Toxics Release Inventory. Much remains to be learned about how to involve citizens effectively in government decisionmaking, but a number of the techniques that were first tried in the context of pollution control provide important lessons and models.

Congress, EPA, the states, and the private sector have developed other tools and approaches that help indicate the path of the future. Market-type mechanisms, particularly sulfur dioxide trading under the 1990 Clean Air Act amendments, have proven to be as effective in practice as in theory, at least under the right circumstances.

EPA, in conjunction with the scientific community, also has made important contributions to the science necessary for dealing with environmental problems. It has advanced the art of conducting risk assessments and pioneered the approach of broad comparative risk assessment. The science of ecology, while still primitive and inadequately supported, has nevertheless benefited from federal and state efforts to understand and deal with pollution problems.

Despite these and other accomplishments, we conclude that the pollution control regulatory system has deep and fundamental flaws. There is a massive dearth of scientific knowledge and data. The system's priorities are wrong, it is ineffective in dealing with many current problems, and it is inefficient and excessively intrusive. Most of the participants in the system are aware of these defects.

There is no consensus about how to remedy these flaws. Not only do disagreements exist among the different interests concerned with pollution control, but even groups that seemingly have a common interest disagree with each other. There is no agreement among large corporations about decentralizing pollution control or about preserving the current regulations. There is no agreement among environmental groups about the utility of market mechanisms.

The United States does not need to wait for a consensus to act on these problems. Our political system is designed to negotiate agreements and find common ground. If we wait for consensus, we will wait forever.

Furthermore, some agreement exists about the principles that should guide changes in pollution control and about the characteristics of a pollution control system for the next century. The future system should be results-oriented, integrated, efficient, participatory, and information-rich.

Results-oriented—The current system is focused largely on how to control pollution rather than on whether pollution is actually being controlled. Technology-based standards have received the most criticism on this score, but ISO14000 (a fashionable industry nostrum that involves an international standard for corporate internal pollution control procedures) is similarly focused on means rather than ends. The system of the future needs constantly to ask whether human health and the natural environment are being adequately protected. Regulators need to set the standards, ensure that adequate data are available to know if the standards are being met, and take compliance measures if the standards are not being met. The means used to achieve the goals are secondary and should largely be left in the hands of the regulated parties.

Integrated—The fragmentation of the current system is a major factor in its lack of rational priorities, its inefficiency, and its difficulty in identifying and dealing with new problems. Within the next decade, most developed nations will have abandoned the medium-oriented system in favor of an integrated approach. The United States should not be saddled with an antiquated and cumbersome approach. An integrated approach, whether based on geographical area, economic sector, function (enforcement, research, standards setting, and so forth), or some combination of these, is a prerequisite to most other basic reforms of the pollution control system.

Interagency or intersectoral integration is a different but equally important challenge. Future environmental quality will be determined by the nation's energy, agricultural, and transportation policies. Better ways to link environmental concerns and these other policy areas need to be instituted.

Efficient—The inefficiency of the current system should no longer be tolerated. Costs should be considered explicitly when establishing goals, and maximum flexibility should be allowed in achieving the goals. The use of market mechanisms should be a priority.

Participatory—Continuing efforts and experiments are required to encourage citizens to have some trust in their government and to participate in the decisionmaking process. However, public participation should not be used as an excuse for government to abandon its role as protector of the public interest.

Information-rich—The current system lacks all kinds of necessary information—scientific and economic information, information about actual environmental conditions (monitoring data), and information about whether programs are working (program evaluation). Recent events and trends—reduced spending for research and monitoring, the dismantling of the congressional Office of Technology Assessment—have made the situation worse. A new system has to recognize the need for information and provide the resources, incentives, and institutions to provide it.

The above characteristics implicitly assume that the federal government will continue to play a central role in controlling pollution and that that role will be based on congressionally enacted laws. Several considerations support a continuing federal role: pollution does not respect state or local borders, many of the most important problems require international action, lack of national standards would interfere with interstate commerce and might well result in a deterioration of environmental protection, and economies of scale exist for such functions as research. We think that law will continue to be centrally important because the kinds of incentives—especially financial gains or penalties—necessary to affect the behavior of potential polluters come primarily through law. Good will by itself is unlikely to move people to action. The question of what the incentives are to change behavior is a necessary question to ask, and the answer frequently leads back to legislation.

The huge gap between agreeing to general characteristics and agreeing to specific changes and initiatives will not be easy to close, but the shortcomings of the existing system are so great that the nation needs to try. Agreement on general principles needs to be followed by the hard work of thinking through detailed policies and negotiating the political compromises necessary to enact and implement them. Failure to make the changes will be costly to the economy, to the environment, and to every citizen.

In this report, we have summarized the strengths and weaknesses of the existing pollution control regulatory system. Our lengthier comprehensive analysis should further help to guide consideration of what changes are needed. The time to start considering these changes is now.